Appalachian Maple

Recipes & Stories from the Mountains

North Mountain Press

Edited by Jeff DeBellis
Illustrated by Hollis Moore
Stories by Kate Fotos

ISBN 978-0-578-75983-8

Published by
North Mountain Press
390 Road Less Traveled
Franklin, WV 26807

Printed by
T&N Printing
205 12th St NE
Charlottesville, VA 22902

Support from the One Foundation and the U.S. Department of Agriculture's (USDA) Agricultural Marketing Service through grant AM170100XXXXG168 made this publication possible. Its contents are solely the responsibility of the authors and do not necessarily represent the official views of the One Foundation or the USDA.

For more information or to order copies, visit www.future.edu.

Introduction

West Virginia is not the first place that comes to mind when most of us think of maple syrup, but with more tappable maple trees than Vermont, maybe it should be. The ways that people collect and use the sap from maple trees is different in West Virginia as well. As is the case with so many things in the mountains, these methods often vary from one valley to the next. This collection shares some of these variations, told by people who collect, cook with, and enjoy maple products from West Virginia and the surrounding mountains of Central Appalachia.

Sugaring - collecting maple sap to make syrup and other products – is experiencing a renaissance in the region. For farmers, it's a way to glean income from land in muddy late winter when other farming activities slow. For chefs, it's a way to showcase one of the many treasures that the bountiful local woods produce. For everyone, it's a delicious natural sweetener.

Appalachian families have been sugaring for hundreds of years, passing tools and techniques from one generation to the next. The earliest known account of sugaring in Appalachia comes from a seventeenth century book called *The History and Present State of Virginia*. The author writes of Native Americans who had been making syrup along the Upper Potomac River "longer than any now living can remember."

The endurance of this annual tradition enables today's resurgence. Many of the people starting or growing sugaring operations in the region remember making syrup with their parents or grandparents. Some continue to use the methods of their forebearers, collecting sap in buckets and boiling it in enormous iron kettles to make just enough to share with friends and family. Others have embraced more modern technology, such as tubing systems, evaporators, and reverse osmosis units. As farmers produce and sell more maple products, chefs and cooks create new ways to use them, often in combination with other local staples such as trout, apples, and venison.

The farmers, chefs, and cooks who contributed to this collection created their recipes using syrup that they or their neighbors produced. Every farm has a story and a terroir that give its syrup a distinct identity. Try a few different kinds and find your favorite. Pretty soon you'll be creating recipes of your own.

Jeff DeBellis

Contents

Primer on Making Appalachian Maple Syrup

Maple trees
Cordless drill
Drill bit
Hammer
Spiles
Buckets w/ lids
Large pot
Jars w/ lids

Identify maple trees while the leaves are still on, then wait for syrup season in early spring, when the days are warm (40°) and the nights are cold (20°). When maple weather rolls around, head out to the maple trees with spiles, buckets, drill, bit (most spiles require a 5/16″ or 7/16″ bit), and hammer.

Drill a 1½-2″ deep hole near chest height in the tree. Be sure to hit healthy wood, avoiding knots, rotting wood, or damaged areas. Gently tap the spile into place. It doesn't need to be pounded hard, but it should be secure.

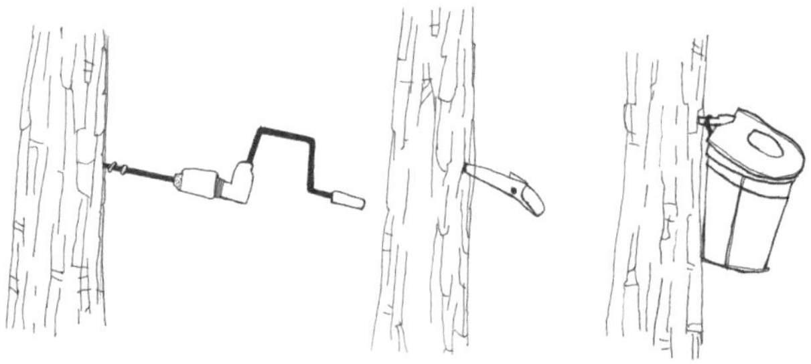

Collect sap frequently. Most people check once a day because sap can spoil. Once sap begins to run, start boiling it down in the pot. Many people do this outside because of the steam that it generates. It takes about 40 gallons of sap to make 1 gallon of syrup. Syrup is done when it is boiling at 7° above the boiling point of water, or when it is thick enough to evenly coat a spoon.

To store the finished syrup, either keep it in the refrigerator or bottle it by heating to 182°, pouring into a clean jar, and topping with a clean lid. The heat will seal the jar.

Kate Fotos
Circleville, WV

Breakfast

Blue Ribbon Scones

These scones have won several blue ribbons at the Helvetia Agricultural Fair - hence the name.

1 cup heavy cream
¼ cup Appalachian maple syrup (plus additional for glaze)
1 cup whole wheat flour
1 cup all-purpose flour
⅓ cup brown sugar
1 Tbs baking powder
½ tsp salt
¼ tsp cinnamon
Confectioners sugar (for glaze)

Preheat oven to 425°. Whisk 1 cup heavy cream with ¼ cup maple syrup. Stir dry and wet ingredients until just blended. Knead against the side of the bowl several times and turn out onto a lightly floured surface. Divide dough in half and form two circles approximately ¾″ thick. Cut each circle into 8 wedges and place on ungreased baking sheet. Bake 12-15 minutes. Make a glaze with maple syrup and confectioners sugar to top scones.

Blue Rock Farm
Blue Rock, WV

Maple Bacon

8 slices smoked bacon

2 oz Appalachian maple syrup

Put bacon in 350° oven until about halfway cooked. During the final minutes of cooking, brush bacon with maple syrup and bake until crisp.

The Greenbrier
White Sulphur Springs, WV

Maple Bacon Cinnamon Rolls

Samantha Hicks' maple bacon cinnamon rolls took second place in the 2019 WV State Fair Savory Maple Contest.

2 large cans cinnamon rolls
1 pkg bacon
1½ cups Appalachian maple syrup
1½ cups chopped nuts

Unroll cinnamon rolls and place on a baking sheet. Place one strip of bacon onto each cinnamon roll. Cover cinnamon rolls with 1 cup maple syrup and roll back into cinnamon roll form. Bake according to package and until golden brown. Remove from oven and cover with more maple syrup. Put the cinnamon rolls back into the oven for about 4-5 minutes. Cover with cinnamon roll topping from package. Top with chopped nuts and remainder of maple syrup.

Samantha Hicks
Fayette County, WV

Maple Walnut Scones

Scones

4 cups all-purpose flour
½ cup granulated sugar
1 tsp baking soda
1 Tbs baking powder
1 tsp salt
½ lb unsalted butter
1 cup walnuts, chopped & roasted
1⅓ cups buttermilk

Mix dry ingredients together in food processor with butter. Pulse until just crumbled texture. With hands, mix in walnuts and buttermilk. Mold into 1″ thick circle, then cut into even triangles. Place on baking tray and bake at 425° until golden brown, about 20-25 minutes.

Glaze

3 cups powdered sugar
½ cup Appalachian maple syrup

Mix powdered sugar with maple syrup, adding more powdered sugar if necessary to make glaze. Glaze warm scones and enjoy.

Dale Hawkins
Fish Hawk Acres
Buckhannon, WV

Deep Roots in Deep Flavors

This story comes from Jeff and Teresa Munn of Loafer's Glory B&B in Franklin, WV.

Maple syrup has long been part of Appalachian life, though the methods for making it have changed significantly over time. In Dahmer, West Virginia, the founding family made maple syrup. They didn't have today's shining stainless steel oil, gas, or wood-fired evaporators. Instead, they used three large cast iron kettles suspended over open fires. Eliza and Jenny Dahmer ladled "sugar water" from one kettle to the next as it began to boil down, making sure not to burn it.

Eliza and Jenny Dahmer boil sap over an open fire in the West Virginia community that still bears their family name.

Without the efficiency of the modern evaporator, sap would sit over the open fire for much longer. How long the sugar in sap caramelizes over heat determines the flavor and color of the syrup. The syrup the Dahmer family made would have been darker and richer in maple flavor than most of the syrup sold today. It would have been almost chocolate-brown in color, with a robust, deep maple flavor, maybe with a touch of smokiness from cooking over an open fire.

Today's Appalachian maple syrup ranges in color from golden yellow to dark brown and tastes anywhere from delicately to robustly maple, but many people in the region still prefer the darker, richer tasting syrups.

The Moonstruck'd Muffin

The creator of this muffin named it for Moonstruck Maple, a syrup producer in Brooks, WV.

Muffins

½ cup butter, room temp
½ cup sugar
½ cup Appalachian maple syrup
2 eggs
1 tsp salt
2 tsp baking powder
2 cups bread flour
1 Tbs lemon zest
½ cup buttermilk
2 cups blueberries

Preheat oven to 375°. Combine butter, sugar, and maple syrup. Add eggs one at a time. Add salt, baking powder, bread flour, and lemon zest. Slowly pour in buttermilk. Softly mix in blueberries and place the resulting batter in greased or lined muffin tins. Place in the heated oven for 30-35 minutes, turning the muffins halfway through. While baking, make streusel topping in a separate and significantly smaller bowl.

Maple Streusel Topping

1 cup butter, cold & grated
½ cup oats
¼ cup Appalachian maple syrup
½ cup flour (add more flour for crumblier streusel)

Add butter, oats, maple syrup, and flour into a bowl. With hands, crumble the mixture. When the muffins have about 10 minutes left to cook, top them with streusel mixture and allow to cook for the remaining time. Once the muffins are done, remove from oven and allow to cool for 10-15 minutes.

Chestnut Revival
Hinton, WV

Orange French Toast w/ Orange Butter Sauce

Orange French Toast

1 loaf French bread

6 eggs

1 cup milk

½ cup orange juice

1 tsp nutmeg

1 tsp cinnamon

2 tsp vanilla extract

¼ cup sugar

Preheat oven to 400°. Slice bread into 1½″ thick slices. Mix all other ingredients together. Soak bread in egg mix until completely absorbed. Arrange bread in a shallow casserole dish coated with cooking spray. Pour remaining liquid into dish. Place dish in oven and bake until all the egg is cooked through, approximately 45 minutes.

Orange Butter Sauce

1 orange

1 cup orange juice

¼ cup sugar

½ cup Appalachian maple syrup

1 stick butter

Grate orange into zest and put aside. Heat juice until reduced by one half. Add sugar, maple syrup, and zest. Remove sauce and add butter. Add extra sugar as desired and serve over French toast.

The Greenbrier
White Sulphur Springs, WV

Vegetables

Apricot Maple Fall Slaw

Slaw

1 purple cabbage, thinly sliced
4 cups Brussel sprouts, roughly chopped
3 cups carrots, slivered
4 oz dried apricots, roughly chopped
½ cup toasted pecans, chopped

Preheat oven to 375°. Combine cabbage, Brussel sprouts, carrots, and apricots in large mixing bowl. Place pecans on baking sheet. Once oven is ready, toast pecans until fragrant, about 10-15 minutes. Remove from oven and allow to cool, then add to the large mixing bowl.

Dressing

¼ cup apple cider vinegar
5 Tbs Appalachian maple syrup
2 Tbs orange juice
½ cup mayonnaise
2 tsp fresh fennel
1 tsp orange zest
1 Tbs garlic, minced

Whisk all dressing ingredients together in a small mixing bowl. Add salt and pepper to personal taste. Pour dressing over the slaw mix. Serve immediately, or cover and refrigerate.

Chestnut Revival
Hinton, WV

Sugarin'-Off in the Bullpasture Valley

Sweet Run Farm's Robin Vance comes from a long line of sugarmakers, starting with his great-grandfather, who was born in 1847. Another notable sugarmaker in his family is his Aunt Alice, who ran the Vance's Country Store. She and her husband Lohr would boil syrup every season in family heirloom kettles, selling syrup and a variety of maple products, like her creamy peanut butter maple sugar blocks. Her operation was also one of the first to support agritourism. A picture of her operation appeared on the cover of the Virginia state map. She also wrote and printed a maple cookbook, *Sugarin'-Off in the Bullpasture Valley*.

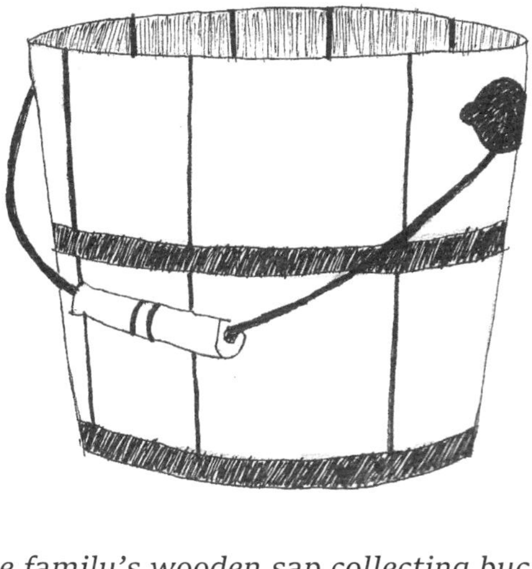

The Vance family's wooden sap collecting bucket, with vine woven hoops.

Family Traditions

Josh Shinaberry is a fifth generation sugarmaker. He operates Cassell's Sugar Shack in Cass, WV.

My family has made maple syrup in the same location near Cass for at least five generations. In the 1970s, my grandfather built the sugar camp I currently use. When I was a child, our entire family would help make syrup. My cousins and I helped gather sap from the buckets. We poured the sap into barrels on a wagon pulled by a tractor that my great uncle drove down a path between the trees. Sometimes we had to remove moths from the buckets that would float on top of the sap.

My grandfather, uncles, and neighbors would sit around the fire talking and feeding it with wood. Grandad would boil syrup all day, shut down the fire before he went to bed, then restart it in the morning. When the syrup was nearly finished they placed a piece of metal roofing over the fire and opened the spigot on the syrup pan.

They poured the syrup into metal buckets and put it into large shallow pans to be "cleared" by mixing a few eggs with some water and putting it into the pans. Once the syrup started boiling we skimmed the top until finally the foam was nearly white. After that we ran it through a wool filter and bottled it. In the spring, people would come to the sugar camp to eat pancakes, buckwheat cakes, and sausage. My grandmother said that as many as 300 people would come to the sugar camp to eat breakfast and buy Cassell's Sugar Shack maple syrup.

After my grandparents were no longer able to make syrup, the sugar camp sat unused for over a decade, until I decided to make syrup there again. I started with plastic milk jugs and some old tubing to collect the sap and slowly worked my way up from there. Now I have about two thousand taps, all on tubing, and still boil sap on the flat pan my grandparents used. I am looking forward to teaching my two young sons the traditions that my grandparents taught me.

Citrus Maple Soba Noodle Salad w/ Crispy Chicken or Roasted Butternut Squash

Enjoy this soba salad recipe on its own or top it with either crispy maple chicken or roasted maple butternut squash.

Soba Salad

1 ten oz package soba noodles (thickness is personal preference)
1 cup toasted coconut
3 Tbs olive oil or other oil of choice
4 Tbs Appalachian maple syrup
½ cup coconut milk
¼ cup tahini
⅓ cup sesame oil
1 orange worth of orange juice and zest
1 lime, juiced
Sea salt
2 Tbs sesame seeds
1 bunch green onion

Preheat oven to 350°. Prepare soba noodles according to package directions. While noodles cook, lay out the coconut on a sheet pan and toss in olive oil or other oil of choice. Place pan in heated oven and cook until toasted, about 10-12 minutes. When noodles are cooked, drain and run cold water over them until they are cool to the touch. In a small bowl, whisk together maple syrup, coconut milk, tahini, sesame oil, orange juice and zest, lime juice, and sea salt. In a larger mixing bowl, combine toasted coconut, sesame seeds, and sliced green onion. Drizzle sauce over noodle mixture and combine well.

Crispy Maple Chicken

4-5 chicken thighs
1 Tbs garlic, minced
2 Tbs ginger, minced
Sea salt
Pepper
¼ cup Appalachian maple syrup
3 Tbs olive oil or grass fed butter

Place raw chicken thighs in a medium saucepan and cover with water. Bring water to a boil and boil for 10 minutes, turning chicken halfway through. Discard the water and lower heat to medium-high. Cut chicken into bite-size pieces while heating in the pan. Add garlic, ginger, salt, pepper, maple syrup, and oil (or butter). Cook for another 10 minutes. Chicken should be cooked through and crispy on the outside.

Roasted Maple Butternut Squash

1 small to medium sized butternut squash
4 Tbs olive oil
⅓ cup Appalachian maple syrup
Sea salt
Pepper

Preheat oven to 375°. Dice butternut squash, add to sheet pan, and toss in olive oil, maple syrup, salt, and pepper. Place pan in heated oven and cook until tender, about 20-25 minutes. Toss squash halfway through to ensure even cooking.

Chestnut Revival
Hinton, WV

Maple Bacon Butternut Squash Purée

1 butternut squash, halved vertically and seeded
1-2 Tbs olive oil, plus more for drizzling
8 slices bacon
2 Tbs butter
½ cup onion, diced
1 bulb garlic, minced
1½ tsp salt
⅓ cup Appalachian maple syrup
⅛ tsp ground nutmeg
Freshly ground black pepper, to taste
4 cups vegetable broth

Preheat oven to 425°. Line baking sheet with parchment paper. Rub olive oil on both halves of squash and sprinkle with salt and pepper. Put squash facedown on parchment paper and roast until tender, about 45-50 minutes. While squash is roasting, cook bacon in a saucepan until done, then remove from pan to drain. Leaving some of the remaining bacon grease in the pan, add butter and onion, then sauté until soft and browning. Add garlic and salt. Cook, stirring often until garlic is fragrant. Put onion and garlic into a blender. After squash has cooled for a bit, use a spoon to scoop it into a bowl, scooping out everything down to the flesh. Add the squash, maple syrup, nutmeg, and black pepper to the blender. Pour in 3 cups vegetable broth and mix. Adjust thickness to personal taste by adding more broth.

Frostmore Farm
Dunmore, WV

Grandfather's Trees

Looking out over Canaan Valley, you notice a beautiful stand of sugar maple trees so large that your arms barely wrap halfway around the trunks. These trees originally belonged to Lindsey Reed, who was a logger by age 14 and a farm owner by 19. Reed cultivated his trees to make maple syrup. He would tap the trees in early spring, then use a sled and horse team to collect all the sap and bring it to his sugar shack. In the 1950s, he replaced his horse team with a crawler tractor, but the product remained the same, a dark and robust maple syrup that Appalachia continues to be known for. Reed saved his sweet syrup for family and friends. His grandchildren all remember him sitting them on a stool in the kitchen, giving them each a piece of maple sugar candy, and sending them home with a quart of syrup.

After Reed passed, the trees stood untapped for years. The sugar shack was eventually torn down and the evaporator pan shoved in the back of the barn, but the memory stayed. Reed's granddaughter, Robin Reed Kalog, now manages the farm. After eyeing the maple trees and dreaming of maple syrup for years, Kalog started small, putting 100 taps into her grandfather's old trees. Kalog hopes to continue her family's tradition of making syrup on the farm and turn it into a wonderful business.

Sweet Potato Salad

Tina Barton of Croagh Patrick Maple Works & Agroforestry made this for a picnic, where most of it got eaten—she took that as a good sign. From Barton: "My husband liked it and said I should make it for us again! Praise indeed! He NEVER wants me to make potato salad unless it is for an event! He doesn't like it with too much mustard, and really doesn't like vinegar. He said the mustard was diluted enough and he didn't complain about the onion."

1 large sweet potato
12-15 baby white potatoes
2 eggs
2 celery stalks
12 baby carrots
1 small yellow onion
¾ cup mayonnaise
1 Tbs yellow mustard
¼ cup milk
1 Tbs Appalachian maple syrup
Salt and pepper to taste

Peel and cube sweet potato. Cube baby white potatoes with skin on. Boil (in salted water) both kinds of potatoes until poke-able with fork. Drain and let cool. Meanwhile, boil eggs about 12 minutes. Let cool. Chop celery, carrots, and onion. (From Barton: *I chopped the onion fairly small. I was worried my husband wouldn't like the raw onion, but decided he often doesn't like potato salad, so he would not be the judge, so I added them.*) Peel and chop eggs once they are cool. Whisk mayo, mustard, milk, maple syrup, salt, and pepper until smooth. Do a taste test and adjust proportions as desired. When the potatoes are cool enough, gently combine the potatoes, egg, mayo mixture, and raw veggies.

Croagh Patrick Maple Works & Agroforestry
Glenville, WV

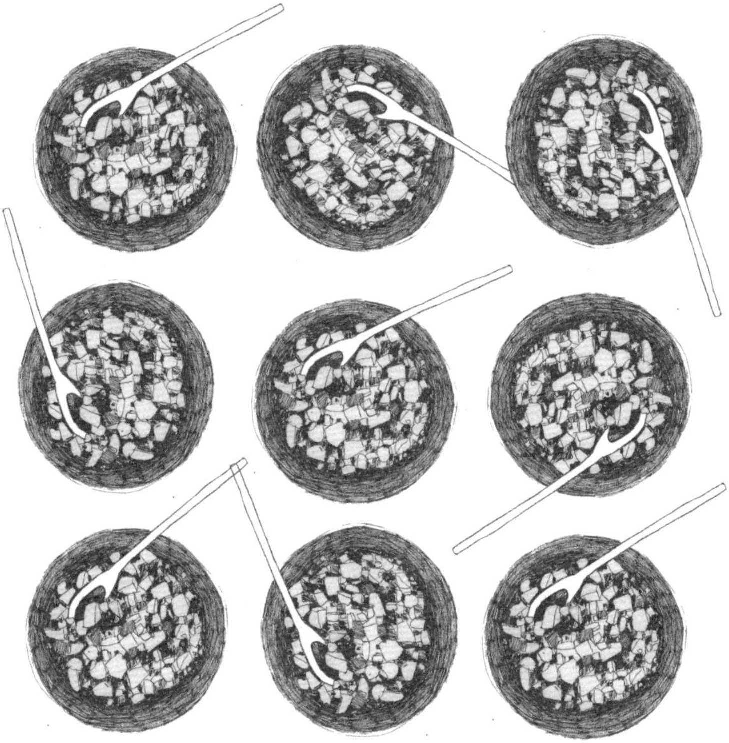

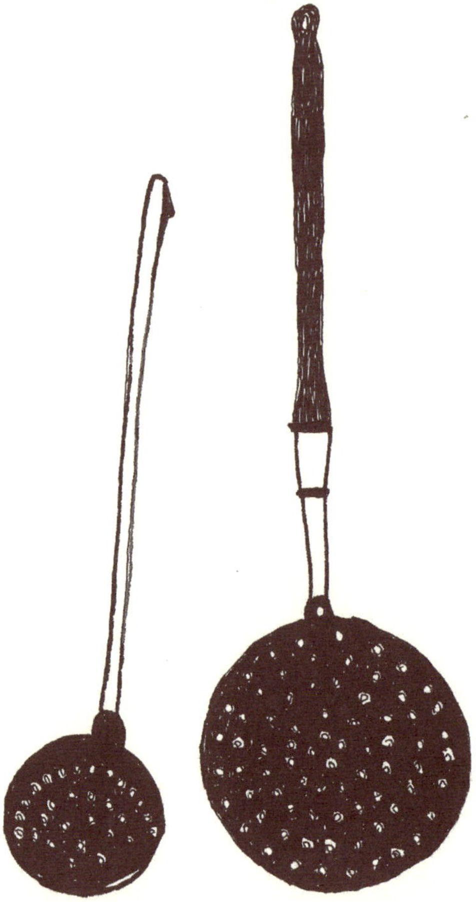

Robin Vance's great-grandfather's kettle skimmer, used for taking the foam off the top of boiling sap. It is over 100 years old. He made it larger and flatter to fit his kettle.

Meat & Fish

Baked Maple Pork Loin

Kathy Hughart's maple pork loin took first place in the 2019 WV State Fair Savory Maple Contest.

1-2 lb pork loin
Flour, enough to coat loin
¼ cup Appalachian maple syrup
¼ cup ketchup
¼ tsp garlic powder

Preheat oven to 350°. Heat oil in skillet. Coat loin in flour, then place in hot skillet, turning meat to brown on all sides. Transfer meat from skillet to deep baking dish. Mix together syrup and ketchup, then pour over loin and sprinkle with garlic powder. Cover and bake in preheated oven until loin reaches 170°, about 45 minutes to 1 hour.

Kathy Hughart
Greenbrier County, WV

Maple Balsamic Baked Salmon

½ cup Appalachian maple syrup
1 Tbs soy sauce
1 Tbs garlic, minced
½ cup balsamic vinegar
1 tsp black pepper
1 lb salmon

Mix maple syrup, soy sauce, garlic, balsamic vinegar, and pepper in a small bowl. Place salmon in a shallow glass baking dish and coat with maple syrup mixture. Cover dish and place in refrigerator to marinate at least 30 minutes, turning once. Preheat oven to 400°. Place baking dish with salmon in preheated oven. Bake uncovered until easily flaked with a fork, about 20 minutes.

Frostmore Farm
Dunmore, WV

Maple BBQ Little Smokies

1 pkg little smokies (or kielbasa)
3 Tbs butter
Appalachian maple sugar, to taste
Ketchup, to taste

Brown little smokies in butter on stove over medium heat. Sprinkle on maple sugar to desired sweetness. Squirt ketchup across pan in a zig-zag pattern. Continue cooking over medium heat, stirring occasionally. Holds well in slow cooker for picnics!

Family Roots Farm
Wellsburg, WV

Maple Glazed Trout

¾ stick butter
16 oz trout fillet
¼ cup Appalachian maple syrup
Salt, to taste

Melt ¼ stick of butter in a pan. Thinly slice the remaining ½ stick. Place trout in pan and top with the sliced butter, then drizzle evenly with maple syrup. Fry until skin is golden brown and crispy. Add salt as desired.

Mountain State Trout
Franklin, WV

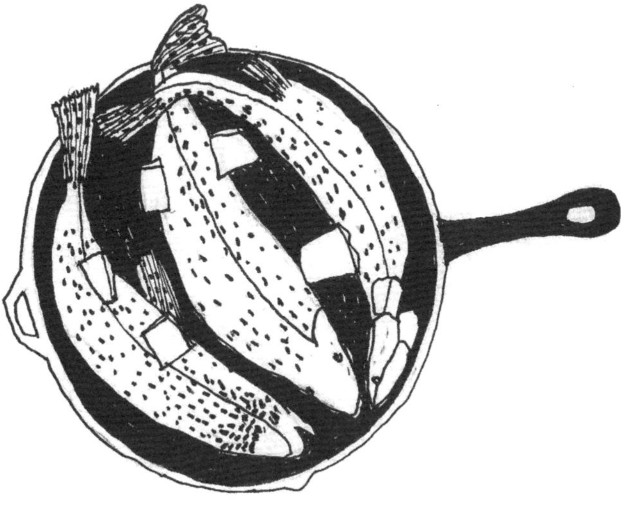

Maple Maryland Crab Cakes

Tina Barton of Croagh Patrick Maple Works & Agroforestry based this recipe on one from Faidley Seafood, an institution at Baltimore's Lexington Market. The original recipe called for saltine crackers and Dijon mustard. She didn't have them on hand once, so she substituted Club® crackers and a honey Dijon and liked the slightly sweet flavor. Some people put Old Bay® in their crab cakes. Barton isn't a fan, and apparently neither was Faidley.

½ cup mayonnaise
1 egg
1 Tbs mustard
1 tsp Appalachian maple syrup
1 Tbs Worcestershire sauce
1 lb jumbo lump crab, with yellow bits and shell picked out
1 cup Club® crackers, crushed
1 Tbs butter
Dash Tabasco® sauce (optional)

Whisk mayo, egg, mustard, maple syrup, and Worcestershire sauce together in a large bowl. Gently stir in crab and crackers, trying not to break up crab lumps. Form into cakes and place on a lightly greased tray. (From Barton: *I don't like non-stick cooking spray because I think it affects the taste. I used softened butter. You pay a lot for that jumbo lump crab. Don't ruin it!*) Let the cakes chill in the refrigerator for a couple hours and they will hold together better. Broil until lightly browned, about 4-5 minutes, flip carefully, and broil 4-5 minutes more until they are toasty brown on top. One fancy Maryland pub serves them on a slice of tomato surrounded by sweet corn. Delicious!

Croagh Patrick Maple Works & Agroforestry
Glenville, WV

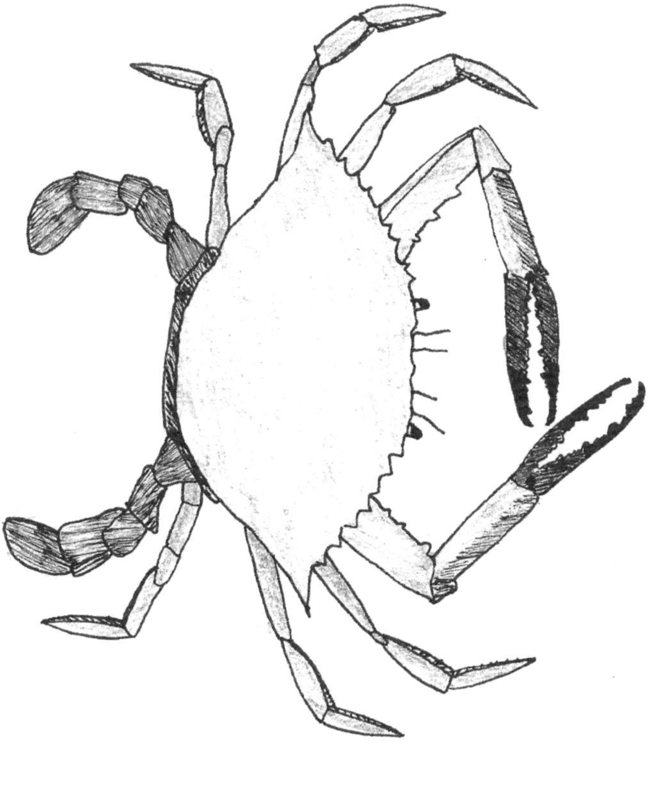

Maple Mustard Kielbasa

¾ cup Appalachian maple syrup
⅓ cup spicy brown mustard
2 pkg kielbasa

Mix maple syrup and mustard in bowl, adjusting to personal taste. Slice kielbasa and brown in skillet. Pour maple mustard mixture onto kielbasa and heat until warm. Goes great with rice!

Heasley Homestead Maple
Bruceton Mills, WV

Maple Venison BBQ

1 cup apple cider vinegar
1 cup Appalachian maple syrup
Red pepper flakes, to taste
3-4 lb venison roast

Put cider vinegar and maple syrup in pot on stove. Adjust the syrup/vinegar ratio to personal taste. Simmer and add pepper flakes to taste. Put roast in slow cooker with half of sauce. Cook on low until very tender, about 8-10 hours. Shred meat. Add remaining sauce. Serve with buns and slaw.

Rosey Santerre
Franklin, WV

Maple Venison Sausage

1 tsp dried thyme
½ tsp dried marjoram
1½ tsp kosher salt
2 Tbs fresh sage, finely chopped (or 2 tsp dried sage)
1 lb ground venison (grind 9 parts venison with 1 part fatty pork for best results)
¼ cup Appalachian maple syrup

Mix dry ingredients in large bowl. Add ground meat and mix well, until herbs are totally incorporated. Add maple syrup and mix well. Form into patties or links. Either fry until done or wrap in foil on a cookie sheet (for easier clean-up) and bake at 350° for 25-30 minutes.

Gillum House Bed & Breakfast
Shinnston, WV

The Sugar Bug

Like many producers in Appalachia, Mike Eye caught the "sugar bug" early, as a young boy in Brandywine, West Virginia. He would help his family carve wooden spiles from sumac and collect sap from buckets. The Eye family would often make enough for several years at a time; only pulling their kettle out when their stores ran low.

As the family grew their operation, they switched from iron kettles to a flat pan with a small furnace that Eye built to use with his father. He eventually built a sugar shack around his furnace and replaced his buckets with plastic tubing.

Eye and a friend now run Jack Mountain Maple. They collect sap from about 600 taps on two properties. They use a reverse osmosis (RO) machine to concentrate the sap, then boil it in a divided flat pan that allows them to draw off finished syrup little by little, instead of in large batches.

Eye's story mirrors those of many other producers across Appalachia. People often start with smaller operations, more for personal use than for sale. As the sugar bug bites, people start tapping more trees, upgrading equipment, and marketing their syrup. Eye, however, stays in touch with his early roots by tapping a few trees on buckets each year.

Peppered Maple Turkey Breast

Turkey

3 Tbs cracked black peppercorns

1 Tbs kosher salt

3-3½ lb boneless, skinless turkey breast

4 oz (1 stick) butter, cut into 8 slices

Preheat oven to 325°. Combine cracked black peppercorns and kosher salt in a shallow, flat container large enough to accommodate turkey breast. Roll turkey breast in peppercorn and salt mixture to coat thoroughly and evenly. Place spice-crusted turkey breast on a baking rack in a roasting pan in preheated oven and cook for 15 minutes. Prepare glaze while turkey cooks.

Maple Syrup Glaze

1 cup Appalachian maple syrup

3 Tbs light brown sugar

1½ Tbs Dijon mustard

Combine all glaze ingredients in a small saucepan and stir until brown sugar dissolves. Remove turkey from oven and brush liberally with glaze. Cook for 5 minutes more, then top with 4 slices of butter. Continue cooking for another 15 minutes, then baste again with maple syrup glaze and cook for 5 minutes more. Place remaining 4 slices of butter on top of turkey breast and continue cooking until a meat thermometer registers 155° at the thickest part of the turkey breast, about another 15 minutes. Remove turkey breast from oven and brush once more with maple syrup glaze. Let rest in a warm place for 15 minutes. Turkey will continue to cook after it is removed from oven and will reach 165° while resting. Pour pan juices into remaining maple syrup glaze and stir to combine. Heat over high heat for approximately 3-5 minutes to reduce slightly. Slice turkey and overlap slices on serving platter. Pour reduced glaze over turkey slices and serve warm or at room temperature.

The Greenbrier
White Sulphur Springs, WV

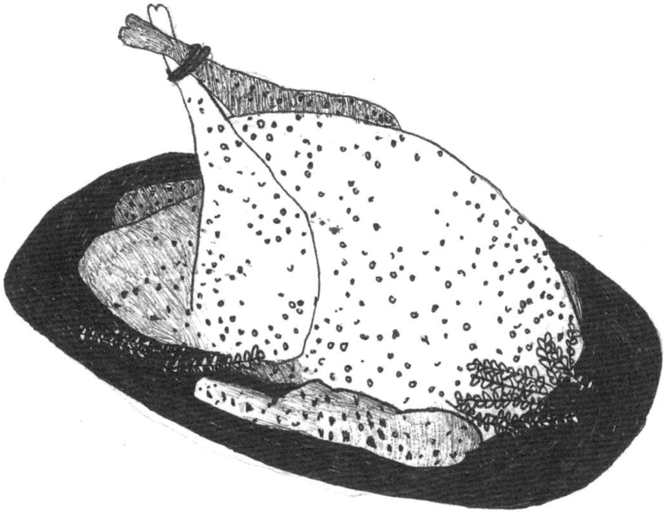

Spicy Maple Glazed Chicken

Diana Norman's spicy maple glazed chicken took third place in the 2019 WV State Fair Savory Maple Contest.

Chicken

10 chicken tenders or wings
1 cup flour
Salt and pepper to season flour
2 eggs
Oil for frying

Roll chicken in seasoned flour, dip in egg, and then again in flour mixture. Fry until golden brown, then drain on paper towel.

Sauce

½ cup Frank's RedHot®
¼ cup Appalachian maple syrup
¼ cup butter
¼ tsp garlic powder

In a small saucepan, mix all ingredients and simmer for 15 minutes. Coat chicken in sauce and serve.

Diana Norman
Greenbrier County, WV

Beverages

Appalachian Maple Apple Spice

8 oz unpasteurized WV apple cider

¾ Tbs grass fed organic butter

1½ oz Smooth Ambler Old Scout® whiskey

1 Tbs Appalachian maple syrup

Apple slice

1 tsp ground cinnamon

1 tsp sugar

Cinnamon sticks

Heat apple cider to a simmer and add butter to melt. Pour whiskey and maple syrup into a mug. Add cider and melted butter mixture and stir vigorously with a spoon. Coat apple slice in cinnamon/sugar mix and add to rim of mug. Drop in cinnamon stick and enjoy!

The Forks Inn
Elkins, WV

Maple Bock Lager

Using straight maple sap in place of water to brew a slightly sweet and heavy beer is a great way to usher in the spring. This works best with a malty beer, such as a bock. Extract kits are available from homebrew supply companies. More experienced or adventurous brewers may choose to forego the kit and modify their own recipe.

5½ gallons Appalachian maple sap
1 bock extract kit
⅔ cup Appalachian maple syrup

Start by bringing 3 gallons of sap to a boil in order to kill any bacteria, then set aside somewhere cold to chill. Looking at the instructions that came with the extract kit, the first step should say to boil about 2½ gallons of water. Use the remaining 2½ gallons of sap instead. Continue following the kit's instructions for the remainder of the boil. When it says to fill a carboy or bucket with cold water, use the 3 chilled gallons from earlier. Follow the kit's instructions for fermenting and lagering the brew. When bottling day arrives, use ⅔ cup maple syrup in place of priming sugar. Cap the bottles and allow the secondary fermentation to take place as indicated by the kit.

Jeff DeBellis
Pendleton County, WV

Maple Whiskey Sour

1½ oz whiskey
1 oz fresh lemon juice
½ oz Appalachian maple syrup

Add all ingredients to an ice-filled shaker. Shake well. Pour over ice and garnish with a lemon peel or wedge.

Karen Milnes
Sweetwater Farm
Sugar Grove, WV

Desserts

Frosted Maple Cookie Bars

Cookie Bars

½ cup butter, softened
½ cup brown sugar
½ cup Appalachian maple sugar
1 egg
2 tsp vanilla
½ cup Appalachian maple syrup
1¾ cups flour
½ tsp baking soda
⅛ tsp salt

Preheat oven to 350° and grease a 9″x 13″ pan. Cream together butter and both sugars on medium speed. Add egg and vanilla, mixing well. Mix in syrup. Sift flour, baking soda, and salt. Add slowly to mixture, mixing well. Spread cookie batter in baking pan and bake 18-20 minutes.

Frosting

1 cup butter, softened
2 cups powdered sugar
¼ cup Appalachian maple syrup
1 tsp vanilla

Beat butter until smooth. Add powdered sugar. Mix on low speed until combined, then switch to high speed for a couple of minutes. Drizzle maple syrup and vanilla while mixing. Mix well.

Family Roots Farm
Wellsburg, WV

The No-Bake Cake

Everett Hedrick began making maple syrup with his family as a young boy. He helped with the woods work, collecting buckets with a horse, sled, and barrel. Once they collected the sap, they boiled it down in iron kettles. In the 1950s, his family moved away from the horse and sled to a pickup truck and water tank to haul the sap out of the woods. Hedrick and his brother continued the family tradition into adulthood, making a dark red maple syrup that had, “no need to flavor or color it.” It was naturally delicious in any form!

One of Hedrick’s early memories of maple syrup is making maple sugarcakes. These cakes don’t require any baking; they made them by continuing to boil maple syrup until it had a higher concentration of sugar, then poured it into forms and let it solidify. Sugarcake makers became so proficient that they could tell the boiling syrup was ready to pour just by the way it dripped off the spoon!

Maple sugarcakes could be stored throughout the year and used like white sugar—except it was cheaper, healthier, and more flavorful. Some Appalachian syrup producers still have sugarcake molds and shavers in their sugar shacks and homes. Hedrick’s family’s secret to combatting the grittiness that often occurs in maple sugarcakes is combining the hot syrup concentrate with a “little bit” of peanut butter before forming the cakes.

Kristin's Maple Dream Cheesecake

Kristin Behrens, a baker at White Grass Café, literally dreamed about this recipe one night while sleeping. She came to work the next day and tried it. It turned out to be an awesome recipe and a favorite at the café!

Crust

1 cup graham cracker crumbs
½ cup finely chopped walnuts
5 Tbs melted butter
1 Tbs sugar
¼ tsp cinnamon

Mix crust ingredients together and press into the bottom of a 10″ springform pan. Wrap springform pan with aluminum foil so that cake can be placed into a pan of water while baking (so water won't leak into cake). Set pan aside and prepare filling. While preparing filling, boil 2 quarts of water to have hot so that the cake can be baked in a pan with water around the cake (about 1″ high). Preheat oven to 350°.

Filling

4 eight oz pkg cream cheese, room temp
⅓ cup sugar
6 eggs
⅔ cup Appalachian maple syrup
½ cup finely chopped walnuts

Beat cream cheese with an electric mixer at high speed until smooth. Add sugar while mixing. Drop mixer to a lower speed and add eggs, one at a time, mixing well. Combine maple syrup and nuts together in a small bowl and then gently stir into cheesecake, mixing as little as possible. Pour cheesecake batter into prepared crust and place cake into a baking pan with 2″ sides, then pour hot water into pan around the cake (NOT ON CAKE) until it reaches up about 1″ on side of cake pan. Bake at 350° for about 50 minutes, until cake is set. Remove cake from hot water and cool, then refrigerate. Serve with Maple Dream Drizzle below.

Maple Dream Drizzle

⅓ cup Appalachian maple syrup
¼ cup finely chopped walnuts

Combine and drizzle over individual pieces of cake before serving.

Kristin Behrens
White Grass Café
Davis, WV

Maple Cheesecake

Filling

8 oz cream cheese, softened

8 oz whipped topping

½ cup Appalachian maple sugar

Blend all filling ingredients together and set aside.

Graham Cracker Pie Crust

1½ cups graham cracker crumbs

⅓ cup Appalachian maple sugar

6 Tbs butter, melted

Blend crust ingredients together and press into 9″ pie pan. Bake crust at 375° for approximately 7 minutes. Cool completely, then add filling. Chill for several hours.

Family Roots Farm
Wellsburg, WV

Maple Chocolate Chip Cookies

1 cup unsalted butter, melted and cooled
1½ cups light brown sugar
½ cup granulated sugar
2 eggs, room temp
1 tsp vanilla
¼ cup Appalachian maple syrup
3¼ cups all-purpose flour
2 tsp corn starch
1 tsp baking powder
1 tsp baking soda
1 tsp salt
1 cup mini chocolate chips
1 cup regular chocolate chips

Stir together butter and sugars in a large bowl. Add eggs, one at a time, stirring until combined. Stir in vanilla and maple syrup. In a separate bowl, whisk together flour, corn starch, baking powder, baking soda, and salt. Gradually add flour mixture to wet ingredients, stirring until completely combined. Stir in chocolate chips. Cover bowl with plastic wrap and allow to chill at least 30 minutes. While the dough is chilling, preheat oven to 350° and prep cookie sheets by lining with parchment paper. Scoop 2-3 Tbs of cookie dough and roll into balls, placing them at least 2″ apart on cookie sheet. Bake about 13 minutes, just until edges begin to turn golden brown. Allow cookies to cool completely on cookie sheet.

Mark & Debbie Bowers
Bowers Maple Farm
Petersburg, WV

From Timber to Syrup

Tina Barton has begun to sugar on her family's farm in Glenville, WV. She started the process of moving their woodlots away from timber production and into forest farming. Many of her recipes, including her pear and maple dishes, are included in this collection.

My grandfather, Patrick W. Gainer, collected folklore and oral traditions of rural life before the days of TV and radio (and of course internet). I never heard him mention maple sugaring, nor is it mentioned in his book, *Witches, Ghosts and Signs*. The only thing he says about sugaring is that most families grew sugar-cane and there would be a person with a sugar press that would go from farm to farm to press the cane, or people would bring their cane to the press. Grandpa inherited land from his ancestors who moved to Tanner (in what is now West Virginia) around 1840. They were farmers, mostly cattle farmers. I have pictures from the early 1900s of the hills, which are so barren! But since then, the forest has returned.

One day about 5 years ago, my sister told me that her husband's family produces maple sugar in New England and she wondered if we had sugar maples on our hill (which used to be Grandpa Gainer's). So I asked my neighbor, Mike Hardman, about it because he is a mountain man if I ever met one, and he knows about things growing in the forest. "These hills are FULL of sugar maples," he said. I promptly forgot about it.

A couple years later, my niece mentioned it again. She shared an article with me about sugaring in West Virginia, and I remembered what my sister and Mike

Hardman had said. So I researched how to identify sugar maples, and, it's true, our hill is full of sugar maples. The only problem is, the land was timbered about 10 or 20 years ago, and most of the trees are too small. However, I found about 15 in the front yard, which was not timbered, and so began my maple production. I found another 15 or 20 trees that escaped timbering and are accessible from a path (I'm 59 and this "bench and break" topography is hard on my old bones!) I tapped 9 trees my first year, then 30, and then last year, 22, choosing the best producers. I only produce a couple gallons of syrup, but I'm proud of what I get, and it tastes better than everyone else's syrup! (Not that I'm bragging!)

One day last fall, Mike Hardman came up the hill chewing on something. "What are you eating?" I asked. "One of your pears," he replied. "I have pears?" I asked, surprised. I had not spent that much time on the hill until I bought my grandfather's house in 2016, and still, I spend most of my time in Maryland, so I have not gotten to know all the good things we have here. These pears are so sweet and crispy, they inspired me to make some pear and maple dishes!

Maple Comfort Pears

This recipe is great as a side dish with pork chops, over ice cream, or just by itself.

16-18 small fresh pears (about 8 cups cubed)
¼ cup Appalachian maple syrup
¼ cup Southern Comfort®

Combine ingredients, bring to boil, cover, turn down to simmer, and cook until soft, about 20-40 minutes (40 minutes for fresh crisp pears). Pears should be soft enough to poke easily with a fork, but not mushy. If they do get mushy, just purée them into a pear sauce!

Croagh Patrick Maple Works & Agroforestry
Glenville, WV

Tin sugarcake molds belonging to the Vance family of McDowell, VA. These molds often turn out better cakes than the more modern rubber molds.

Maple Glazed Pecans

Laurel Fork Sapsuckers held a dessert contest as part of an educational workshop they hosted. This recipe from Kevin and Emily Conner of nearby Mill Gap Farms was one of the winners.

¼ cup Appalachian maple sugar
2 Tbs Appalachian maple syrup
1 Tbs water
2 cups pecans
1 tsp vanilla

Pour maple sugar, maple syrup, and water into a large skillet. Turn on medium-high heat. The higher the heat, the faster the glaze cooks. Mix ingredients with a silicone spatula until liquefied. Add pecans and begin stirring. After pecans are coated with the delicious syrup liquid, add vanilla by dripping around pan. Continue stirring until small dry crystals form on pecans. Once pecans turn a light tan they are ready to spread out on foil or wax paper. As tempting as it may be, let them cool 5-10 minutes before eating. Store in a resealable bag - will last 1-2 weeks in refrigerator.

Kevin & Emily Conner
Mill Gap Farms
Monterey, VA

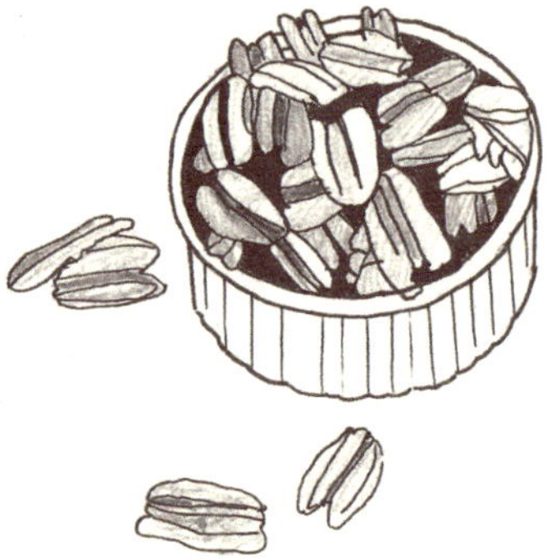

Maple Icebox Pie

Crust

1 pkg graham crackers

½ cup pecans

1 stick butter

Use food processor or high-quality blender to mix pecans and graham crackers. Melt butter and stir into mix. Mold into bottom of pie plate or dish.

Filling

½ lb Appalachian maple cream

8 oz cream cheese, room temp

12 oz tub whipped topping

Mix all filling ingredients together in a blender and pour into crust. Put into freezer until it sets. Serve chilled and enjoy immediately.

Frostmore Farm
Dunmore, WV

Forgotten Syrup

It is not uncommon in Appalachia to hear of forgotten jars of maple syrup hiding in dark corners of cellars, barns, and garages, gathering dust until someone new moves in or spring cleaning rolls around. Maple syrup, when bottled correctly, is shelf stable, but how old is some of this forgotten syrup?

In Lincoln County, West Virginia, a homeowner found a bottle of maple syrup in his garage with only a partial label. Dusting off the bottle and squinting at the faded text, he saw it was dated 1859. With little idea where the bottle came from, the homeowner decided to take it to his local gas station, where he saw Ronk Family Farm's maple syrup, handed the bottle to the clerk and told him to give it to whoever sells them syrup.

The clerk quickly called Paul Ronk and told him to come down immediately. Ronk hopped in his car, drove down to the gas station, and inspected the bottle. Besides the torn label, the bottle looked perfect, full of syrup, with small sugar crystals at the bottom.

"That's when I knew the syrup was still good," Ronk said. With the help of another maple expert, he was able to extract some of the syrup to test the sugar content and the taste. It was still sweet and tasty. "It didn't taste quite like maple, but more butterscotch," Ronk explained. Though no one is sure where the syrup came from, Ronk thinks it came from Lincoln County, or at least West Virginia. The bottle is now on display at the Ronk Family Farm, and may start traveling around to local historical societies and museums.

Maple Ice Cream

4 eggs
1 cup white sugar
1 cup Appalachian maple sugar
5 cups milk
4 cups half n' half
4½ tsp vanilla
½ tsp salt

Mix all ingredients, pour into 1 gallon freezer bag, and freeze.

Ronk Family Farm
Alum Creek, WV

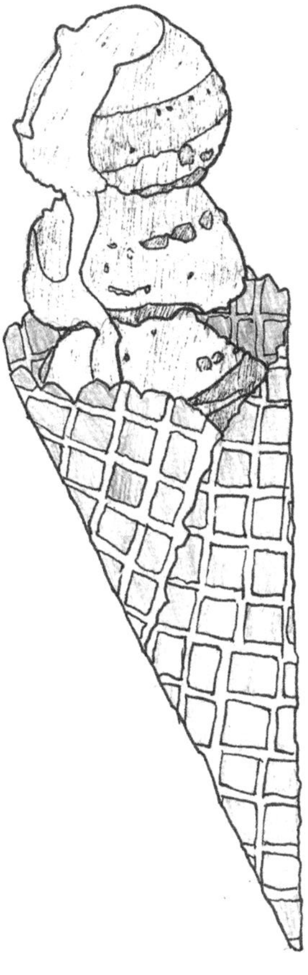

Maple Pecan Crème Brûlée

This recipe, from Chef Thomas Brown, is a favorite and standard at the Guesthouse Lost River.

1 quart heavy cream
½ tsp salt
10 large egg yolks
½ cup Appalachian maple syrup
½ tsp vanilla extract
½ cup turbinado sugar (raw sugar)
Pecan halves for garnish

Preheat oven to 350°. Combine cream and salt in heavy bottom saucepan and place on medium heat. While cream is heating, combine egg yolks, maple syrup, and vanilla extract in a large mixing bowl. Whisk to combine thoroughly. The mixture should drip smoothly from the whisk, with little to no streaks.

Once the cream comes to a boil, remove from heat. With a ladle, carefully add small amounts of cream to the egg mixture while whisking constantly. Do this about 3 or 4 times to temper the egg mixture, and avoid cooking them too soon. Carefully add the remainder of the hot cream to the egg mixture while continuing to whisk. Once the cream is fully incorporated, continue mixing for about a minute.

Pass the mixture through a fine mesh sieve and transfer contents to a pourable vessel (e.g. water pitcher or liquid measuring cup with spout). Place porcelain ramekins into a deep roasting pan. Pour the warm cream mixture into each ramekin, filling them almost completely. If there are bubbles sitting on top of the mixture, use a spoon to hold them back while pouring into the ramekins.

Once ramekins are filled, carefully transfer roasting pan into oven. Before closing, fill the pan with warm water until it reaches about halfway up the ramekin. Carefully slide the pan into the oven. Bake for 40 minutes. Once finished, the custards will jiggle very slightly, but will not appear liquid. If they slosh around too much, they aren't finished. If they don't move at all, they're likely overcooked. If using a thermometer to check the temperature of the custards, they should register between 170° and 180°. Once the crème brûlée has finished baking, remove from water bath and allow to cool at room temperature for about 30 minutes. Place in refrigerator for 2-3 hours, or overnight.

To serve, spread about 2 tsp of turbinado sugar evenly over each crème brûlée. Tilt the ramekin around to help distribute the sugar. Using a gas kitchen torch, melt and caramelize the sugar, until golden. Top with pecans and serve immediately.

Thomas Brown
Guesthouse Lost River
Lost River, WV

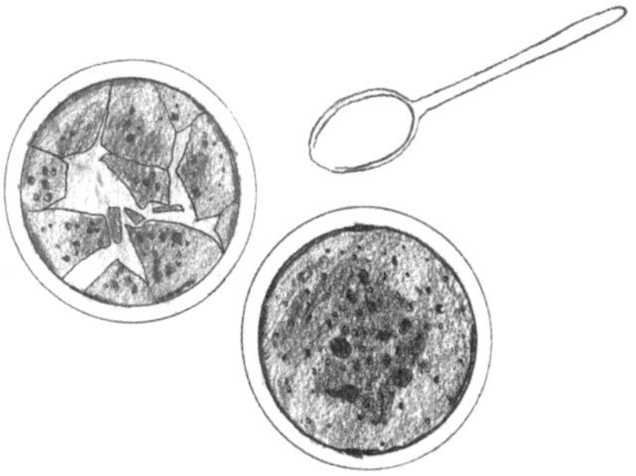

Maple Walnut Blonde Brownies

Cake

1⅓ cups walnuts, chopped
11 Tbs unsalted butter
½ cup Appalachian maple sugar
½ cup Appalachian maple syrup
1 tsp salt
1 egg
1 tsp vanilla extract
2 cups all-purpose flour
¼ tsp baking soda

Preheat oven to 375°. Line a 9″x 9″ baking pan with parchment paper. Add nuts in one layer. Toast walnuts in oven 5-7 minutes. Let cool. Melt butter in a small saucepan over low-medium heat. Stir in maple sugar and cook 2-3 minutes, stirring often until the sugar is mostly melted. Stir in maple syrup. Remove from heat. Pour mixture into a large mixing bowl. Let cool. Stir in salt, egg, and vanilla by hand. Stir in flour and baking soda. Stir in 1 cup of toasted walnuts. Spoon into pan and spread evenly. Bake 22-25 minutes, until lightly golden and mostly set in the center—a slight jiggle when shaking the pan is okay.

Icing

1 Tbs Appalachian maple syrup
½ tsp vanilla
1 cup confectioners sugar

Pour maple syrup into small bowl. Whisk in vanilla and confectioners sugar. Add more sugar if needed to make a thin glaze. Drizzle over cake and sprinkle remaining toasted walnuts over the top.

Family Roots Farm
Wellsburg, WV

Pear & Maple Pie

Tina Barton of Croagh Patrick Maple Works & Agroforestry created this pie using the small, crispy pears that grow on her property and chestnuts from her neighbor's property in Gilmer County, WV. Once roasted, chestnuts have the texture of an underdone, plain baked potato.

Filling

6 cups pears, peeled & sliced
12 chestnuts, roasted (optional)
¼ cup Appalachian maple syrup
1 tsp cinnamon
2 Tbs flour
1 pinch nutmeg
1 Tbs butter

Mix pears with chestnuts (if desired). Separately, combine syrup, cinnamon, flour, and nutmeg. Drizzle maple syrup mixture over pears and chestnuts. Stir to coat, adding more syrup or sugar if desired.

Crust

2¼ cups all-purpose flour
1 tsp salt
½ cup vegetable oil
6 Tbs milk

Combine all ingredients and mix with a fork. Form into two balls. Sprinkle water on pastry rolling surface. Put down a 12″ square of wax paper. Place one of the dough balls on the wax paper and flatten slightly. Cover with another sheet of wax paper. Roll out dough between the sheets until it forms a rough circle, about

12″ in diameter. Carefully remove top layer of wax paper, lift bottom layer with dough on top, and flip onto a pie plate. Center dough and remove wax paper. Put pear mixture into crust and dot with butter. Roll out second dough ball in same way, flip it into place on top, remove wax paper, crimp edges, and poke holes to vent. Bake at 375° for 40-45 minutes.

Croagh Patrick Maple Works & Agroforestry
Glenville, WV

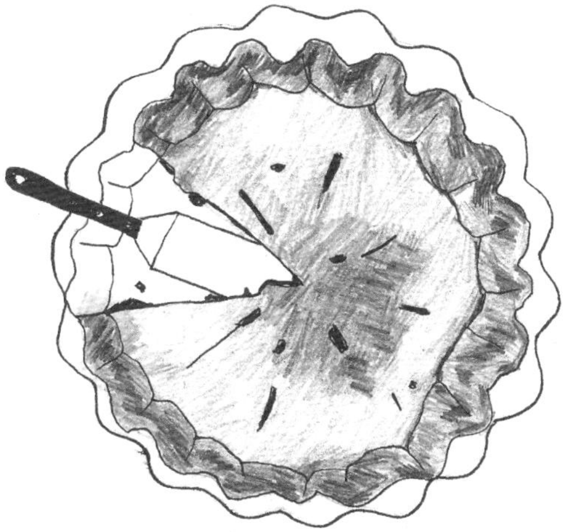

Sugar on Snow

Appalachian maple syrup
1 pan hard-packed snow or crushed ice

Heat maple syrup to 235°. Pour hot syrup over snow or ice. Allow to set up and enjoy wrapped onto a popsicle stick or spoon.

Kate Fotos
Circleville, WV

We are deeply grateful to all of the contributors, as well as advisers Keith Heasley, Karen Milnes, Mike Rechlin, and the following organizations:

Experience Learning, based in Circleville, WV, uses adventures and the outdoors to help people realize their individual and collective potential. They team up with the WV Maple Syrup Producers Association to run Maple in the Classroom, which connects schools with nearby maple producers to teach the art and science of maple production through demonstrations and experiments. **www.experience-learning.org**

Future Generations University, based in Pendleton County, WV, has been a pacesetter in promoting research, learning, and action toward inclusive and sustainable community change since it was founded in 1992. Future Generations' Appalachian Program conducts research and training to help create income opportunities, while promoting land stewardship. **www.future.edu**

West Virginia Maple Syrup Producers Association promotes, educates, and researches maple and tree syrup, as well as value-added syrup products, throughout West Virginia. Most of their member farms focus on maple syrup and related products, while some are exploring other tree syrups such as sycamore and walnut. The association helps West Virginia maple syrup producers connect to share tips about tapping trees, where to get good equipment, and other experiences. They also share information with farmers interested in starting sugarhouses. **www.wvmspa.org**

Index

Notes

Notes

Notes